Jan Solonar
Dmitry Reznick

Results of experimental studies of electrical contact

Jan Solonar
Dmitry Reznick

Results of experimental studies of electrical contact

ScienciaScripts

Imprint
Any brand names and product names mentioned in this book are subject to trademark, brand or patent protection and are trademarks or registered trademarks of their respective holders. The use of brand names, product names, common names, trade names, product descriptions etc. even without a particular marking in this work is in no way to be construed to mean that such names may be regarded as unrestricted in respect of trademark and brand protection legislation and could thus be used by anyone.

Cover image: www.ingimage.com

This book is a translation from the original published under ISBN 978-3-659-84924-4.

Publisher:
Sciencia Scripts
is a trademark of
Dodo Books Indian Ocean Ltd. and OmniScriptum S.R.L publishing group

120 High Road, East Finchley, London, N2 9ED, United Kingdom
Str. Armeneasca 28/1, office 1, Chisinau MD-2012, Republic of Moldova, Europe
Printed at: see last page
ISBN: 978-620-6-88386-9

Gian Solonar, Dmitry Reznik

Experimental results of moving electric contact of unipolar machines

CONTENTS.

1. Analysis of electric power sources

As sources of DC electrical energy in power installations, conventional collector DC generators, alternators with rectifiers, and unipolar generators (UGs) can be used.

The use of superconducting windings will increase the density of electrical energy in these machines and reduce their specific weight, which is associated with an increase in magnetic flux in the working volume and a reduction in heat losses.

However, it is known that in collector electric machines there is a dependence of the maximum allowable magnetic field of the armature on the magnetic field of the inductor.

The strong magnetic field of armature reaction in synchronous machines causes them to go out of synchronism during transients.

In DC machines, the strong magnetic field of the armature makes it difficult for the collector to operate.

In general cases, the main disadvantage is that a fairly large loss of generator voltage is required to reverse the armature current.

It should also be noted that in alternators, an alternating magnetic field acts on the superconducting winding.

In addition, the winding is subjected to rotational torque from the rotor side and must therefore be connected to the machine housing, which increases the power required to cool *it.*

Compared to other types of electric machines, unipolar generators have a number of advantages. Simplicity of design, high overload capacity, high efficiency, absence of ripples in current and voltage curve, possibility of direct connection to the turbine of power unit, etc.

The use of a superconducting winding in unipolar generators makes it possible to significantly increase the magnetic field and the generator efficiency. In this case, the superconducting field winding is placed in a fixed cryostat and is not exposed to the alternating component of the magnetic field or any mechanical effects, as in other electrical machines. The operating characteristics of unipolar generators are largely determined by the linear rotor speed.

To a wider application of UG in power installations, is not so much the strength characteristics of the rotor material, as the ability to stable operation of the movable electrical contact. At present, brush and liquid-metal movable electric contacts are mainly used in unipolar machines.

Brush contacts in stationary mode can only operate up to speeds of 20-30 m/s at an electric current density of $2 \cdot 10^5$ A/m^2 .

In short-term mode, when using carbon-fiber brushes with copper coating, a linear speed of rotation up to 150 m/s was achieved at a current density of 10^5 A/m^2 .In addition to mechanical losses in the contact zone there are also significant electrical losses. And the increase of rotor speed leads to the increase of electrical losses.

The use of liquid-metal contact (LMC) is a more effective solution to the contact issue in unipolar generators.However, such contact can operate only up to 50m/sec, because at higher rotational speeds the contact ring becomes unstable and losses due to friction increase significantly.In addition, in a strong magnetic field, the breakage of LMC is possible due to the pushing of metal out of the interelectrode gap by electromagnetic forces, as a result of which, an electric arc is formed in the zone of metal absence, disturbing the normal p

In addition to LC and brush contacts, electric arc contact can also be used in unipolar generators.The voltage drop in the electric arc of such contact reaches 1218 V at electric current density up to 110A/cm^2 .As a result, in the combustion zone, the arc generates large heat flows, which makes it difficult to cool the electrodes and leads to a significant decrease in the efficiency of the generator.It should be noted that the rotor speed of 150m/s achieved in these tests did not affect the characteristics and stability of the electric arc.As additional studies have shown, this type of contact can operate up to 350 m/s, and under creation of certain conditions, the voltage drop in the contact arc can be reduced to 1-3 V at current density up to 100 -200 A/cm^2 .Achievement of such rotor speeds, and losses in the contact make it possible to consider the unipolar generator as a highly efficient source of electric energy directly connected to the turbine, which is also one of the positive factors in the creation of power units of large capacities.

2. Experimental stands

With the assistance of scientists of NIITP, Moscow, Dr. A.A. Porotnikov, Dr. I.N. Ostretsov, Dr. A.N. Ageyev, in the seventies, an experimental unit was created in the NIITP branch, which was used to investigate the possibility of using high-speed plasma contact in unipolar machines. These studies confirmed the assumptions of scientists.

In the eighties the works were continued with the support and financing of the scientist of Kharkov Institute of Physics and Technology (KIPT), Kharkov, Dr. N.A. Khizhnyak.

When creating experimental stands and conducting research on this topic, Solonar D. took active part. P., Panasyuk M.A., Goloviev Yu.

In the Kremenchug branch of KHPI experimental stands were created, where the work on research of plasma contact of unipolar machines was continued.

The following requirements have been taken into account.

Stands should provide the possibility of creating physical conditions in the contact similar to those of unipolar machines, which are characterized by such parameters:

- electric current, in the contact arc 0 - 2000 A;
- magnetic field produced by the field winding 0 - 2000 ersted;
- preliminary vacuum in the working volume of the product 10^{-2} -10^{-3} mmHg;
- temperature of the experimental product 570 K;
- providing a cesium vapor pressure of 10^{-2} mm Hg.

In addition, visual inspection and recording of such physical parameters is necessary during the experiments:

- the amount of electric current flowing in the arc of the electrical contact;
- voltage drop in the plasma contact arc;
- voltage drop on the experimental item:
- the value of current flowing through the field winding of the SD;
- of the field winding voltage;
- magnetic field strength in the contact arc zone.

Control over the value of electric current and voltage at different parts of the electrical circuits of the experimental product and installation, as well as the magnetic field strength was carried out by means of standard electrical measuring devices.
All instruments were placed on the control panel of the experimental setup and connected to the connectors of the measurement panels installed at the workstations.
The experimental setup where the research was conducted consisted of two stands: stand 1 and stand 2.
Stand 1, Fig. 2.1, was intended for experimental studies of a unipolar motor with a thermoemission current-carrying unit.

Figure 2.1 - Stand 1 for experimental studies of unipolar engine

Stand 2 was intended for studies of cathode assemblies (CA) and preliminary studies of cathode assembly design in the low-voltage arc burning mode with fixed electrodes and without magnetic field.
Stands 1 and 2 had a common power supply system containing nine rectifiers with an output power of 10 kW each and two rectifiers having smooth regulation of the output voltage in the range of 3-12V at a rated current load of 3200A. Besides, the stands are equipped with autonomous vacuum systems consisting of forevacuum pumps and diffusion pumps. A common control panel is used to control the stands, which allows remote control of the system operation 6

power supply and vacuum systems, which makes it possible to change the necessary physical parameters and modes of operation of the research object during experiments.

Figure 2.2 - Experimental stand for the study of cathode assemblies.

At the control panel were mounted control and measuring and recording instruments, connected by autonomous measuring lines to the stands, which together with the primary transducers of measured values make up the measurement system of the experimental unit.

An experimental unipolar motor was installed on stand 1. Structurally, the UD was made of two main parts:

1. of the field winding in the form of a solenoid;
2. vacuumized casing in which the rotor of the SD and the cathode assembly were installed.

The solenoid forming the excitation winding consisted of two independent housings mounted on movable carts. Inside the housings there was a winding, in the form of galettes, with geometric characteristics $\alpha = 1.4$ and $\beta = 1.1$ with an inner diameter of 0.45m and a length of 1.0m. This geometry of the excitation winding, with the available power of the power supply system, made it possible to obtain the magnetic field strength in the working volume of the SD of about 1300 ersted.

The rotor housing of the SD was a cylindrical chamber with an inner diameter of 0.39 m, and consisted of a running and flange parts. In the body of the running part, a shaft was mounted on rolling bearings, to which a disk, which is the rotor of the SD, was attached. On the flange part, on its axis was installed an insulated electrode, on which the cathode units were attached. The inner cavity in this electrode served as a channel

for caesium vapor supply to the working volume of the SD.
For visual observation of the operation of cathode units and electrical contact of the SD, there was a viewing window in the case.
The required temperature regime of the inner surface of the working volume of the SD was provided by a heater made of nichrome tape installed on the outer side of the flange part of the casing.
The measurement lines were discharged from the working volume of the SD by means of a multichannel vacuum-tight connector.
The unipolar motor was connected to the vacuum system by means of a special pipe on the flange side.
All flange joints of the case were sealed with vacuum rubber and had water cooling, which ensured tightness of the case when it was heated up to the temperature of 500-700K.
To conduct preliminary studies of experimental cathode assemblies (CA), a special vacuum chamber installed on bench 2, Fig. 2, was designed and manufactured. The chamber design allowed to reproduce the necessary physical conditions realized in the working volume of the DC. Geometric parameters of the chamber (inner diameter 0.39 m) make it possible to test various CHP designs, up to full-size ones, with their further installation in the SD.
In addition, the design of the chamber allowed, if necessary, to install it inside the solenoids of the stand 1 and conduct experimental studies in the magnetic field.
The design of the experimental slipstream assembly consisted of the following main parts:

- cathode, solid or hollow, made entirely or only on the working surface of a material with high emissive properties, e.g. tungsten;
- cathode unit housing designed for fixing the cathode, heater and installation of the unit at the workplace.

The unipolar motor was heated by means of an ohmic heater made of nichrome tape and located on the cylindrical part of the housing.

The saturated vapor pressure of cesium was determined by the lowest temperature of the inner surface of the working volume.

Before the UD heating was switched on, a vacuum of 10^{-2} torr was reached in its working volume, which after heating to the operating temperature of 570--620°K deteriorated and did not exceed 510^{-2} torr.

During experimental studies, the temperature of the cathode and anode was measured.

The temperature of the anode at a stationary disk was measured at the point located on the opposite side of the disk from the cathode, in the burning zone of the electric arc contact.

The cathode temperature was measured with a thermocouple welded to one of the sector bodies on which the cathodes were attached.

Before breaking the cesium ampoule, after heating the product to the required temperature, its working volume was disconnected from the vacuum system.

3. Experimental study of a current-carrying device with hollow cathodes

In the seventies, experimental studies of movable plasma contact in a unipolar motor were carried out at the Moscow branch of NIITP.

Tungsten plate cathodes were used in the studies, which made it possible to obtain in the electric arc contact at current up to 1200 A and density up to 200 A/cm^2 voltage in the arc contact up to 2-4 V. Based on these studies, it was concluded that it is possible to create unipolar electric machines (motors and generators) with such a contact.

However, when discussing these studies, some doubts were raised about the stability of PC operation. One of these doubts concerned the stability of the circular electric arc of the contact, both without and in the magnetic field.

In addition, NIITP scientists hypothesized that there may be negligible losses when creating electrical contact operating conditions similar to thermal emission converters in PCs

To verify this issue, five studies of the current-carrying assembly were carried out. Due to the lack of developed designs of cathode assemblies and refractory materials, tungsten plates one mm thick were used as cathodes.

In tests 1-4, cathode assemblies were used, which consisted of two coaxially arranged tungsten plates 1mm thick 9

forming a hollow cathode. The gap between the plates was set by spacers made of tungsten plates 0.5 mm thick. The end working surface of such a cathode, with a plate width of 30 mm and thickness of 1 mm, was about 0.6 cm^2 .The cathodes were fixed on a support ring with a diameter of 300 mm, made of 18N10T.
In the first four tests, 11, 10, 7, 12 cathodes were used, respectively, and in the fifth test, 32 tungsten plates bent downward were fixed on the ring. The working thermoemissive surface of such a cathode, formed by bending its leading edge, was 1.5 cm^2 , and the total thermoemissive surface of the cathode assembly reached 48 cm^2
The study of the characteristics of the electric arc contact when using these slip assemblies was carried out on the experimental stand No. 2 ,
Preliminary vacuum in the working volume of the SD, before cesium vapor injection, did not exceed $6 \cdot 10^{-2}$ mm Hg.
The temperature of the coldest inner surface of the product reached 600K. The temperature of the cesium ampoule body was 620 - 650K. The temperature of the cathodes was 600K. The interelectrode gap between the anode (disk) and cathodes was 2.5-3 mm.
After applying voltage to the electrodes (cathode and anode), in the presence of cesium vapor in the working volume of the SD, an electric arc of contact appeared first on one or two cathodes and gradually, in the process of heating the cathodes, spread to the other cathodes. After some time, not more than 3060s, the cathodes were heated up to the thermoemission temperature. In the viewing window, after this time interval, the arc was observed from all cathodes. When the voltage was switched off and switched on again, the contact arc started to burn from all cathodes.
At the same time, a luminous plasma halo was also observed on the side surfaces of the cathodes, although the intensity of its luminescence was much less than in the interelectrode gap
With decreasing electric current, this plasma halo shifted toward the end part of the cathodes, and when the current increased, it again spread over most of the current surface of the cathodes.
As shown by observation through the viewing window IN THE fifth test, where a ring current collector was used, the arc burned steadily across the ring from the cathode ends
Table 3.1 shows the results of studies 1.1-1.5,

Table 3.1 - Results of experimental studies of plasma contact without magnetic field

Electric current in the PC arc, A	Voltage drop in PC, V	Research conditions
1	2	3
Study 1		
700	4,0	T_X = 570K; T_{FL} = 630K; T_{CS} = 650K; T_{ENG} = 810K; T_{EC} = 1120K; $l_{ПК}$ = 3mm; t = 10 min. A cathode assembly with 11 cathodes with a total working surface S = 6.6 cm was used in the tests[2] The current density in this study was 60-200A/cm^2
800	3,5	
480	3,0	
400	3,0	
600	3,5	
700	4,0	
400	3,0	
700	4,0	
600	3,2	
1400	4,5	
800	4,5	
400	3,0	
800	4,0	
Study 2		
1000	4,0	T_X = 590K; T_{FL} = 630K; T_{CS} = 620K; T_{ENG} = 770K; $T_{ЭК}$ = 1180K; $l_{πк}$ = 3mm; t = 5 min. A cathode assembly having 10 hollow cathodes with a total working surface of S = 6.0 s m^2 . was used in the studies. The current density in this study was 30-330A/cm^2
800	4,5	
600	5,5	
800	4,5	
700	5,5	
300	3,2	
200	2,8	
400	2,9	
600	3,2	
800	7,0	
2000	5,5	

Continuation of Table 3.1

1	2	3
	Study 3	
100	1,65	TX = 570K; TFL = 620K; TCS = 620K; TC = 730K; $T_{ЭК}$ = 1000K; $l_{пк}$ = 3mm; t = 7 min. A cathode assembly. having 7 hollow cathodes, with a total working surface S = 4.0·s m^2 Current density in this study was 25 - 150A/cm was used in this study^2
200	2,8	
300	2,4	
400	2,3	
500	3,0	
200	1,7	
600	3,0	
200	1,6	
	Study 4	
300	1,9	TX = 690K; TFL = 720K; TCS = 690K; TEN = 720K; $T_{ЭК}$ = 900K; $1_{ПК}$ = 3mm; t = 8 min A cathode assembly with 12 hollow cathodes with a total working surface of S = 7.2 cm^2 was used in the studies. The current density in this study was 50-100A/cm^2
400	2,8	
600	3,8	
800	4,5	
200	2,0	
400	3,0	
600	3,8	
180	2,0	
400	3,4	
200	2,0	
400	3,2	
	Study 5	
2000	2,5	TX = 600K; TFL = 650K; TCS = 600K; TENG = 700K; $T_{ЭК}$ = 900K; $1_{ПК}$ = 3mm; t = 4.5 min A cathode assembly having 32 flat cathodes with a total working surface of S = 48.0 cm^2 was used in the studies. The current density in this study was 40-70A/cm^2
2100	2,2	
2400	3,0	
2600	3,0	
3000	3,5	
3400	3,5	
2000	3,3	

The magnitude of current in tests 1.1-1.4 varied from 100 to 2000 A., the voltage drop in these tests reached 1.65-5.5 V.

In the fifth test, the electric current in the arc was varied from 2000 to 3400 A, and with the total thermionic surface area of the cathode assembly equal to 48.0 cm^2 , the calculated current density was 40-70 A/cm^2 , respectively.

The contact arc voltage drop in this study was as high as 2.53.5 V, which agreed well with the results of Tests 1.1-1.4.

Figure 3.1 also shows the voltampere characteristics of these studies.

The duration of the tests was 5-10 min.

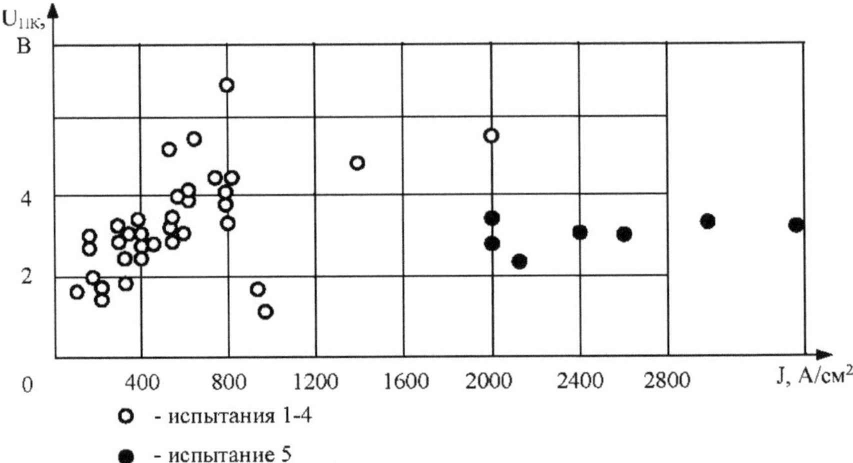

Figure 3.1 - Voltampere characteristic of electric arc of plasma contact in studies 1.1-1.5

After testing, the slip ring and anode were in good condition.

The current density in the electric arc was determined on the basis of the calculated thermoemission surface of the cathodes, which in the first four tests was taken equal to the end area of the ETU cathodes, and in the fifth test - to the end surface of 32 tungsten plates.

The dependence of the voltage drop on the electric current density is shown in Fig. 3.2. Since, the electric arc contact was observed over the entire surface of the current collector cathodes, the calculated current density in the electric arc contact was assumed to be 30-350 A/cm^2 .

After testing, the slip ring and anode were in good condition.

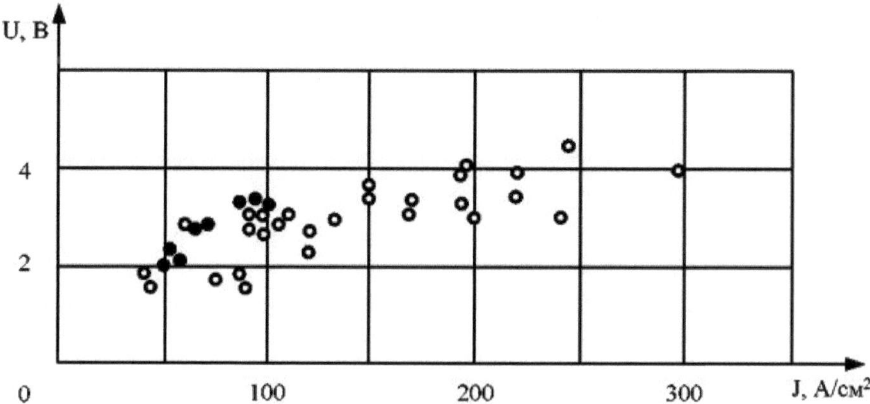

Figure 3.2 - Voltampere characteristic of electric arc of plasma contact in studies 1.1-1.5

4. Experimental investigations of plate sliding device in unipolar motor

To investigate the plasma contact in the magnetic field of a unipolar motor in the operating mode, an experimental bench was set up to study different models of cathode and their operation in a longitudinal magnetic field.

The plasma contact was investigated when used as a movable electrical contact in a unipolar motor, on the experimental bench shown in Fig. 2.2.

To remove electric current from the shaft, a liquid-metal contact was used, the working body of which was a gallium-indium alloy.

The current-carrying assembly was recruited from four tungsten plates used in the fifth test, fixed on one ring with a diameter of 300 mm. The working thermoemissive surface of such a cathode, formed by bending its leading edge, was 1.5 cm^2 , and the total thermoemissive surface of such a cathode assembly reached 6 cm^2

The plate cathodes were heated by an electric arc burning between the main and heating cathode, which was also recruited from four tungsten plates arranged perpendicularly to the main cathode

After heating the inner surfaces of the working volume up to 580 K and the ampoule body up to 600 K and disconnecting it from the vacuum system, the ampoule was broken and cesium vapor filled the UD volume. When voltage was applied to the main heating cathode, an electric arc appeared, heating the main cathode to a temperature of 1200-1300 K.

As observations through the viewing window showed, the electric heating arc burned uniformly over the entire surface of the heating cathode. The magnitude of the current flowing in the heating arc reached 320 A, and the voltage drop decreased as the cathodes heated up from 6.5 to 5.0 V.

After heating the main cathode, a voltage was applied to the anode, disk, and cathode and, between them, a moving contact arc was generated, which burned steadily mainly from all plate cathode ends.

Plasma contact studies were carried out both without magnetic field and in a longitudinal magnetic field. The results of the study are summarized in Table 4.1 and as a voltammetric characteristic in Fig. 4.1.

Table 4.1. - Results of experimental studies 2.1-2.2 of plasma contact in a magnetic field

Electric current in the arc PC A	Voltage drop in PC, V	Magnetic field in the PC zone, E	Research conditions
1	2	3	4
Study 2. 1			
400	5	0	T_X = 560K; T_{FL} = 600K; 1∏K = 3mm; t = 10 min. A current-carrying assembly having four plate cathodes with a total working surface of S = 6·cm^2 . was used in the tests. The current density in this study was 70-250A/cm^2 The rotor speed was varied in the range of 0 - 5000 rpm.
400	4,5	0	
500	4,1	0	
600	3,5	0	
700	3,0	0	
650	3,3	400	
500	4,3	1300	
800	5,7	1300	
1400	3,1	1300	
1500	3,0	1300	
1600	2,5	0	
1400	2,6	0	
1600	2,7	0	
600	2,75	0	

Continuation of Table 4.1

1	2	3	4
Study 2. 2.			
200	3,0	0	TX = 560K; TFL = 600K; IPK = 2mm; t = 10 min. A current-carrying assembly having four plate cathodes with a total working surface of S = 6 cm^2 . was used in the tests. The current density in this study was 30-330A/cm^2 .
400	3,25	0	
600	3,5	0	
400	4,0	0	
700	2,5	400	
700	2,5	600	
700	2,5	800	
600	2,7	1200	
1000	3,0	1200	
1000	3,5	400	
1300	3,7	1300	Rotational speed of the rotor changed in 0 - 7000 rpm, or 115m/sec.
1400	3,5	0	
2000	3,3	0	
1300	3,5	800	
800	2,8	0	
700	2,8	0	
700	2,7	0	
1300	3,8	1200	
2200	3,0	0	

When the magnetic field was increased to 1300 ersted, the glow on the side surfaces of the main cathode plates disappeared and the discharge burned only from the end surfaces.

Two tests were carried out. As follows from Table 4.1 and voltampere characteristic at change of electric current in the range of 200-2200A the voltage in the contact arc varied, in the range of 2.5V - 3.8V

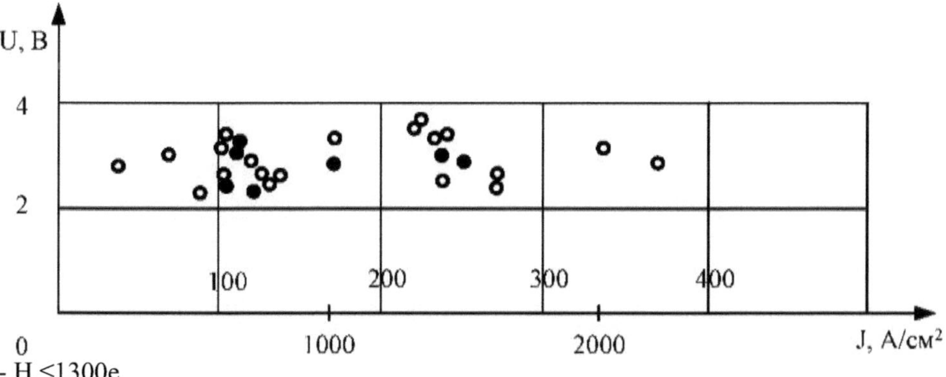

Figure 4.1 - Effects of magnetic field on the voltammetric characteristics of the electric arc of the plasma contact in studies21-2.2.

When the magnetic field was increased to 1300 ersted, the glow on the side surfaces of the main cathode plates disappeared and the discharge burned only from the end surfaces.

Two tests were carried out. As follows from Table 4.1 and the voltampere characteristic when changing the electric current in the range of 200-2200A, the voltage in the contact arc varied, in the range of 2.5 V - 3.8 V

Increasing the magnetic field to 1300 ersted did not significantly increase the contact voltage, Figure 4.1.

Rotation of the disk at a speed of 0-7000 rpm, which corresponded to a linear velocity in the electric arc zone of the moving contact of 0-115 m/s, had no noticeable effect on the characteristics and stability of the contact arc.

After testing, the slip ring and disk were in good condition with no signs of melting, The duration of the tests was up to 10 min, and was determined by the heating of the UD rotor.

5. Experimental studies of a current-carrying device with multi-cavity cathodes

Experimental studies were also carried out on the unipolar motor bench,

The constructional scheme of the slip current collector is shown in Fig. 5.1 The cathodes of this current-carrying device were made of 5 tungsten plates 40 mm wide and 0.3 mm thick, compressed in a package between two plates of steel Kh18N10T, The thermoemissive surface of such a cathode was 0.6 cm^2 .

Three cathodes each were mounted on the support sectors. The thermoemissive surface of the sector cathodes was about 2 cm^2 .The sectors were fixed to the current supply ring

Five experimental studies 3.1-3.5 were conducted using such current slip devices.

In these tests, the preliminary vacuum before the introduction of cesium vapor, when heating the inner surfaces of the working volume of the SD to a temperature of 600-610 K, did not exceed 510^{-2} mm Hg. The initial temperature of the cathodes at filling the working volume of the SD with cesium vapor did not exceed 700 K.

In the three tests 3.1, 3.2, 3.3 the slip assembly was used, consisting of six cathodes having a thermionic emissive surface area of 4cm^2 , and in Tests 3.4 and 3.5 the current collector assembly had six sectors of three cathodes each with a total area of 12.5cm^2 .

Figure 5.1 - Design scheme of the slip ring device

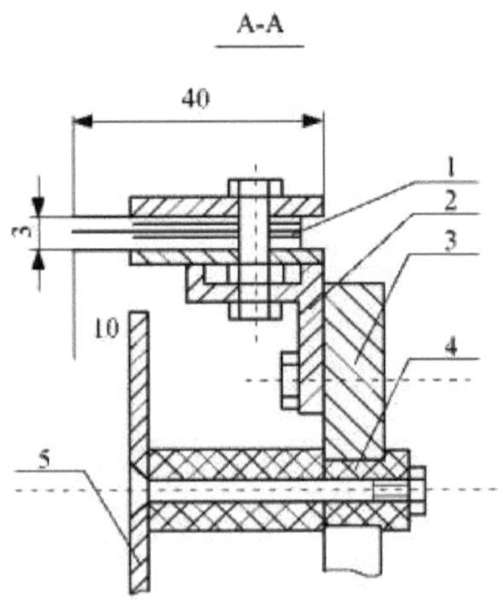

After heating the unipolar motor to a temperature of 600-610 K and reaching a caesium vapor pressure of 510^{-2} tor its working volume was disconnected from the vacuum system and filled with caesium vapor.

When voltage was applied to the slip assembly and anode (disk), an electric arc was generated between them. During the initial period in tests 3.1, 3.2, 3.3, unstable burning of the arc from different places of the cathodes of the slip assembly was observed through the viewing window. At first it burned from one to two cathodes and then spread to the other six cathodes. In the future, the arc burning was stable and steady.

At the value of the electric current equal to 150 - 200 A, with the current density from the cathode equal to 100-280 A/cm^2 , the arc burned only from the end surfaces of the cathodes. With increasing current, the arc spread to the side surfaces of the cathodes. The most intensive arc burning was observed from the cathode cavities.

When using six, sectors with a total area of 12.5 cm^2 , the electric current density was 12-16A/cm^2 , and the intensity of electric arc burning from the cathode ends was different, which could be caused by different temperatures of the cathode surfaces and unequal interelectrode gaps.

At magnetic field strength equal to 1300 ersted, the discharge burned only from the end surfaces of the cathodes. In addition, rotation of the disk at a speed of 6000-7000 rpm or 115 m/s had no appreciable effect on the parameters of the contact arc.

When the longitudinal magnetic field was switched on, disk rotation occurred. Moreover, the increase of the magnetic led to a decrease in the area of arc burning from the side surfaces of the cathodes and to a more stable arc burning without a noticeable deterioration of the arc parameters.

The most characteristic results of experimental studies are presented in Fig. 5.2 in the form of voltampere characteristics. As can be seen from these voltammetric characteristics, all experimental points are practically on two VACs, which is due to the different current density in the PC. In studies 3.1, 3.2, 3.3, the value of current varied from 100 to 800 A. and the voltage drop in the contact arc was 3-6 V. At the same time, the electric current density in the contact arc at the thermoemissive surface of cathodes equal to 4cm^2 , reached 25-200A/cm^2 , and the voltage drop in the contact arc was 3-6 V.

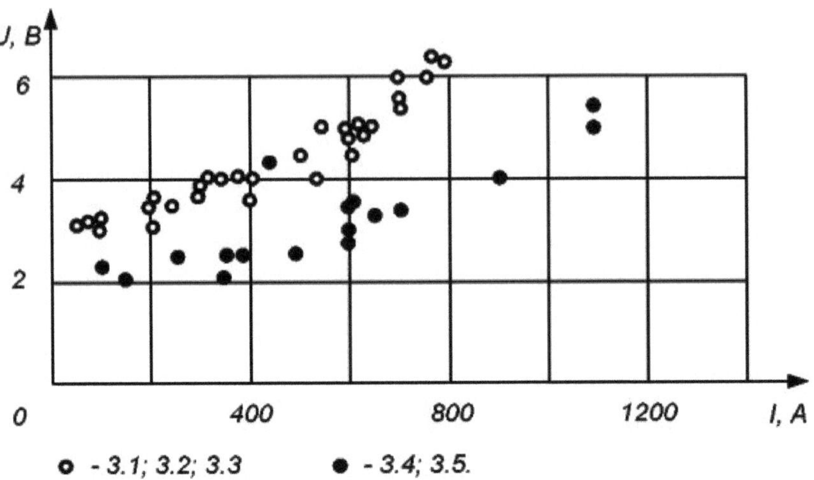

Figure 5.2 - Voltampere characteristic in studies 3.1-3.3 of electric arc of plasma contact

In studies 3.4, 3.5, the current magnitude was increased up to 1100 A, at an electric current density of -100A/cm^2 .and the voltage drop in the PC arc did not exceed 5 V. If we analyze these tests by the current density of Fig. 5.3, within the error made in determining the current density, the VACs of all tests coincide.

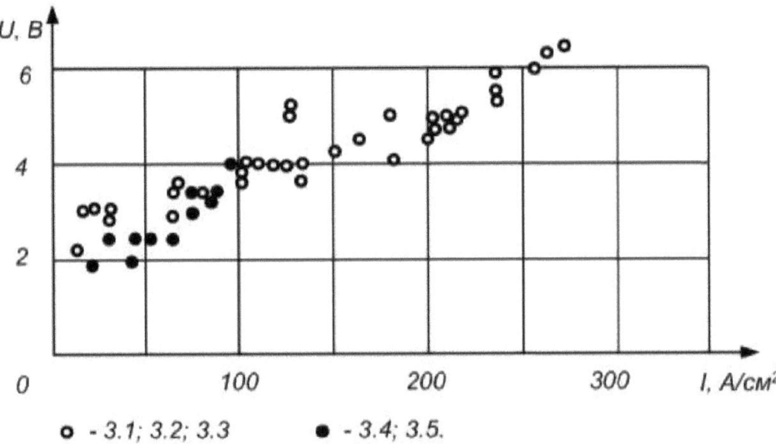

Figure 5.3 - Voltampere characteristic of the electric arc of the plasma contact in studies 3.1-3.3.

6. Results of experimental investigations of the sliding current collector with cesium supply to the contact zone

In tests 4.1. - 4.5 multi-cavity cathodes were used, made of tungsten plates 0.5 mm thick and 40 mm wide (Fig. 6.1). Between the plates at both current ends along the entire length were spacers made of tungsten plates 0.5 mm thick.

After welding the plates and between each other and welding them to the collector 2, a cathode with cavities between the plates was obtained. The cathodes were installed in the current-supplying electrode, into the cylindrical cavity of which caesium was poured from the ampoule body. On the outer surface of the electrode there was a heater designed to heat the cesium to the required temperature.

Cesium vapor was fed through the spigot and collector openings of the cathode into its cavities and further into the combustion zone of the contact arc.

Four cathodes were fixed in the current-supplying electrode. The gap between the cathodes and anode was 1 mm and 3.5-4 mm.

Before feeding caesium into the inner cavity of the current-supplying electrode, the barocamera was heated to 570 - 620 K, and a vacuum of 2 - 6 10^{-2} mm Hg was reached in its working volume.

After caesium vapor supply into the electrode cavity with the baro-chamber valve closed, a holding time of 1 min was performed to stabilize caesium vapor supply into the contact zone.

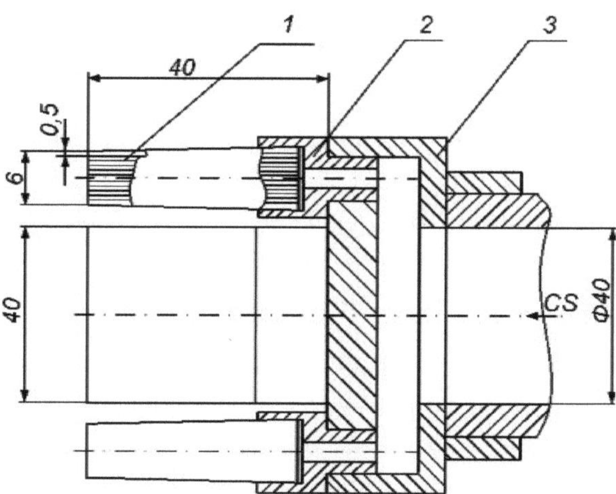

Figure 6.1 - Multi-cavity cathode assembly with cavities between plates

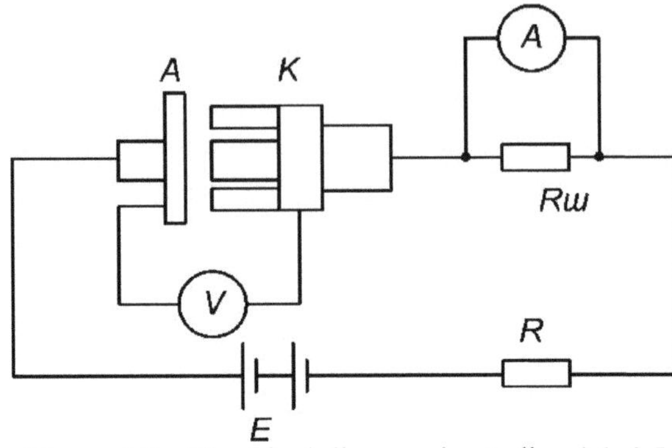

Figure 6.2 - Electrical diagram in studies 4.1-4.5

Although the temperature of the cathodes at the beginning of the tests did not exceed 630 K, however, when voltage was applied to the electrodes (provided that cesium vapor was present in the contact zone between them, a discharge occurred. At first it burned unsteadily from different places of the cathodes, and then it stabilized and burned only from the ends of the cathodes. At the same time, the greatest intensity of discharge was observed in the cavities of the cathodes.

Moreover, with increasing electric current through the contact, the intensity of the electric arc glow in the gap increased, and at a current exceeding 250 - 300 A, the arc burned also from the side surfaces of the cathodes. However, the intensity of arc burning in the cavities of the cathodes was much greater than from other places of the cathodes.

When the voltage was turned on again, the arc was ignited only from the end surfaces of the cathodes.

At the same time, the cathode temperature did not exceed 1270 - 1370 K and the anode temperature did not exceed 770 - 870 K.

The test results are presented in the form of voltammetric characteristics in Fig. 6.3.

If we do not take into account the results of the initial, unstable modes of combustion of the electric arc, the VAC has an almost straight-line dependence in the range of current 150 - 500 A. At such values of electric current, the voltage in the contact arc varied from 2.0 to 4.5 V, although there are experimental points lying slightly below this voltampere characteristic. Moreover, the value of the interelectrode distance did not have a significant effect on the characteristics of the contact electric arc

After the tests, the electrodes were in good condition. The duration of the tests was 10 - 12 min.

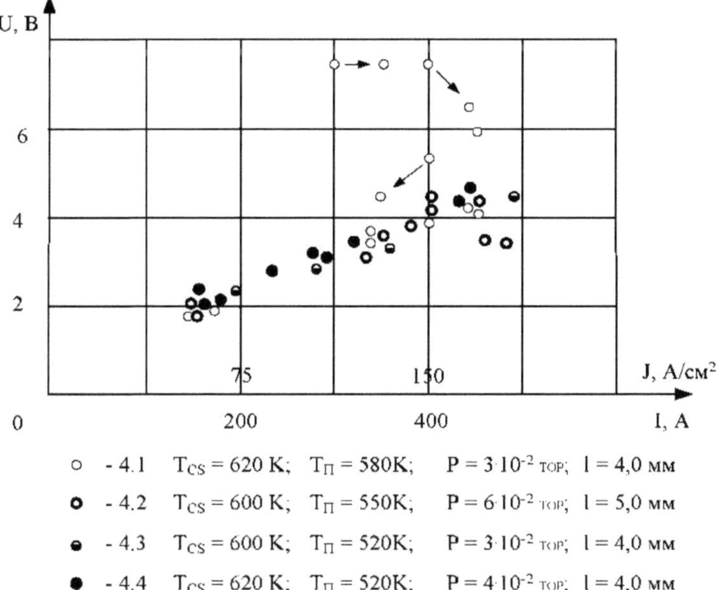

Figure 6.3 - Voltampere characteristic of the electric plasma contact in studies 4.1-4.4.

However, as can be seen from Fig. 6.4, where the results of tests 4.5, 4.6, 4.7 are presented, the size of the gap between the anode and cathode significantly affects the WAC and especially it is noticed when the current is increased.

As shown by the imprints of the electric arc on the anode and visual observations in all tests, the arc burned not over the entire end surface of each cathode, but only from a part of it, which was about 0. 5- 0.7 of this surfaces. In this regard, the assumed current density in the contact arc at these current values was 75 - 200 A/cm^2 .

In Test 4.5, in which the gap was left at 3.5 mm, the voltage drop in the contact arc also varied in the range of 2.0 to 4.5 Although the temperature of the cathodes at the beginning of the tests did not exceed 630 K, however, when voltage was applied to the electrodes (provided that cesium vapor was present in the contact zone), a discharge occurred between them. At first it burned unsteadily from different places of the cathodes, and then it stabilized and burned only from the cathode ends. The greatest intensity of discharge was observed in the cavities of the cathodes.

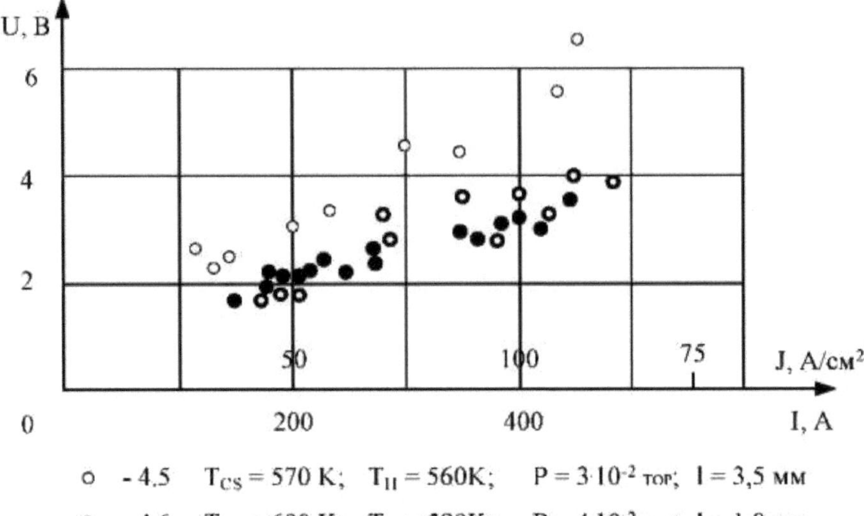

Figure 6.4 - Voltampere characteristic of the electric arc of the plasma contact in studies 4.5.-4.7.

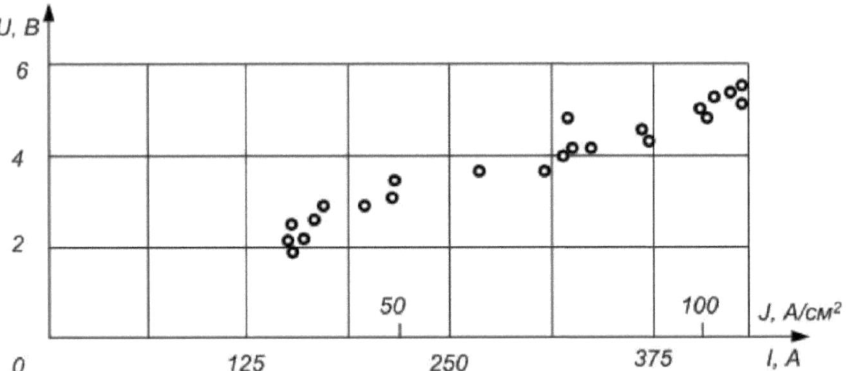

Figure 6.5 - Voltampere characteristic of electric arc of plasma contact in study 4.5

In addition, the arc burned also from the side surfaces of the cathodes, although the intensity of its burning from these surfaces was much less than from the cavities. Moreover, the area of the surface on which the glow was visible varied with the change in the current value. At a current less than 200 A, the electric arc burned practically only from the cathode ends.

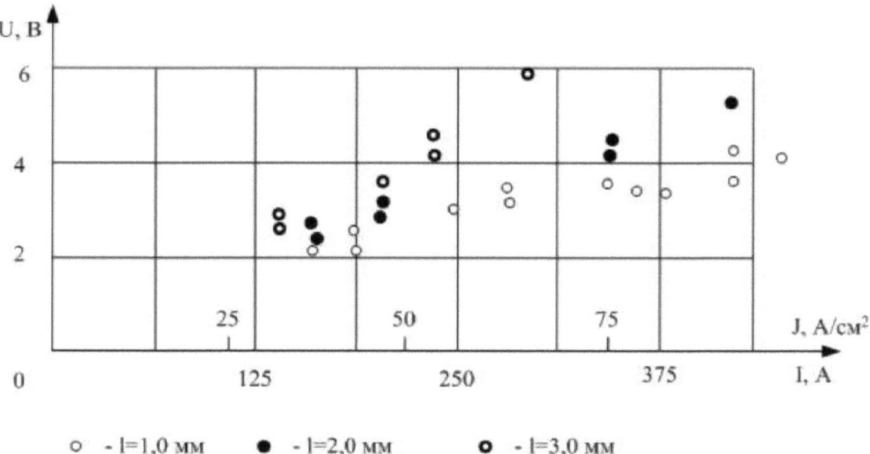

Figure 6.6 - Voltampere characteristic of the electric arc of the plasma contact in Study 4.8.

During the study 4.8, the gap was varied and at gaps of 1mm, 2mm and 3mm, the WACs were taken.

As the gap between the cathode and anode increased, the arc voltage also increased. This was especially noticeable at currents greater than 250 A.

As can be seen from the results of studies at a gap equal to 1mm minimum voltage in the arc varied from 1.8 to 4 V with an increase in current from 160 to 480 A.

At a gap equal to 3.5 mm, the arc voltage at the same current values was noticeably higher and varied from 2.8 to 6.5 V.

If we take into account that mainly the electric arc burned from the cavities of the cathodes and at that not along their entire length, the calculated current density in the arc was 50 - 120 A/cm^2 .

After testing, the electrodes were in good condition.

The duration of the tests was up to 10 min.

7. Studies of plasma contact with electric arc heating of the cathode

In the cathode unit KU-2, an additional electric arc burning parallel to the main magnetic field of the UM was used to heat the working surface.

The cathode assembly KU-2 had three main cathodes, each of which consisted of two parts of cylinder I and body 5, Fig. 7.1.At the end of the cathodes, there were five holes of 1mm diameter each, through which cesium vapors passed into the discharge zone. The main cathodes were inserted tightly into the holes of the collector body and fixed in it

Figure 7.1 - Experimental cathode assembly with arc heating

The collector was designed to bring caesium into the cavity of the main cathodes, where it vaporized and entered the discharge zone.

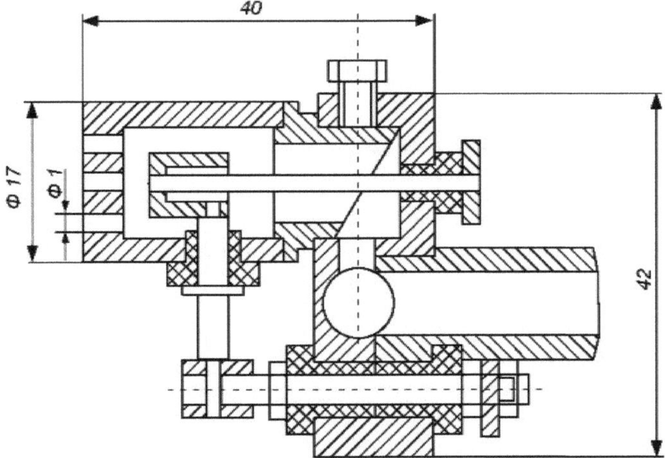

In the cavities of the main cathodes were located, heating cathodes in the form of rods, isolated from the main cathodes by bushings from ABN.

The cathode body I had three blind holes of *ϕ12*, in which the heating cathodes electrically insulated from the cathode body were located. The cavities in the cathode body were connected by holes with a diameter of 6 mm.

The heating cathodes were attached to the collector through electrically insulated bushings made of ABN 2 and connected to each other by a current conducting plate 3.

Cesium vapor from the current-supplying electrode, to which the cathode assemblies were attached, flowed through holes in the cathode ends into the interelectrode gap, i.e. into the electric arc of the contact.

Heating of the cathode end surface was carried out by an electric arc burning between the heating cathode and the main cathode.

In addition, it was assumed that the heating arc would also produce ionization in the main contact arc, which should result in reduced PC arc losses.

Tests of this cathode assembly were conducted to investigate the performance of its design, the possibility and feasibility of using this cathode assembly.

In the experimental studies, the electric current flowing in the electric arcs of the heating cathode and contact arc was measured, as well as the voltages in the electric arcs between the heating cathodes and the main cathodes, and between the anode and the main cathodes The electrical scheme of the measurements is shown in Fig. 7.2.

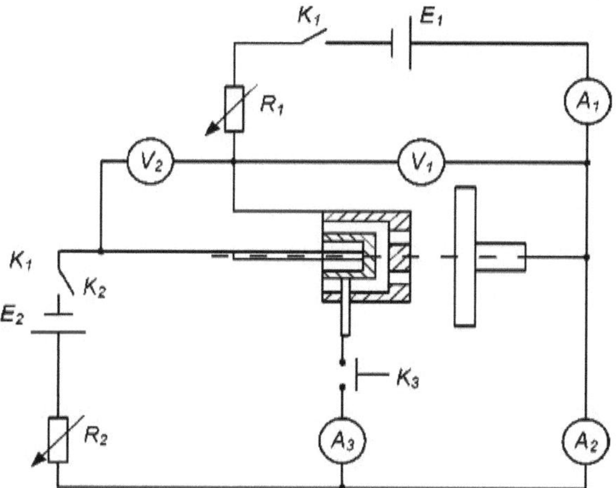

Figure 7.2 - Electrical and measurement diagrams in studies 6.1-6.6

In Study 6.1, a cathode assembly with three cathodes was installed in preparation for testing, Fig. 7.1. The gap between the cathodes and the anode was 3 mm, and the gap between the heating cathodes (HC) and the main cathodes was 2 mm.

After heating the cathode assembly body and vaporizer to 600 K, the working volume of the product was disconnected from the vacuum system.

At first, the cathode heating circuit by contactors K_3 and K_2 was connected to the heating source E_2 (Fig. 7.2) and when current flowed through the heating rods and cathodes, they were heated up.

An ampoule of cesium was broken and liquid cesium was poured into the cavity of the current-supplying electrode, where the cesium vapor evaporated and entered the electrical contact zone.

After the heating circuit was broken by the contactor к3. and when cesium vapor was fed into the cavities of the cathodes, an electric arc occurred between the heating cathodes and the main cathodes, as well as the anode, disk.
The viewing window showed that at first the arc was burning in the cavities of the three cathodes and from the holes on the cathode ends. The heating current reached 110A at a voltage drop between the heating cathode and anode equal to 6 V, the voltage drop between the main cathodes and anode was 2.2 V, Table 7.1. As a result, their rapid heating occurred.
When the source E_1 was switched on, an electric arc appeared between the cathodes and anode, disk. The magnitude of the total current in the electric arc contact between the cathode and anode increased from 110 A to 260-300 A, and the voltage drop $and_{нагр}$ increased by only 2+ 3 V. The voltage drop between the main cathode and anode did not change and was 2.2 V, Table. The discharge burned steadily for 3 minutes The study was stopped due to the possibility of anode melting.
The second study was also conducted using three cathodes, At first, the cathode heating circuit by contactors K_3 and K_2 was connected to the heating source E_2 (Fig. 7.2) and when current flowed through the heating cathodes, they were heated up.
After heating the heating cathodes and applying voltage to the electrodes and cesium vapor to the discharge zone, a heating arc occurred, resulting in heating of the cathodes. The magnitude of the current when the source E_2 , was switched on varied from 60 A to 200 A (Table 7.1), and the voltage $and_{нагр}$ from 5, 8 V to 8 V.
After 100 - 200 sec, the heating arc burned from all holes on the cathode end. There was a rapid heating of the working surfaces of the cathodes.
When only the heating arc was burning, the voltage between the cathodes and the disk did not exceed 2.2V.
In the observation window it was seen that at first the arc burned steadily, in the cavities of three cathodes and from five holes at a constant value of current reaching 125 A and voltage drop in the interelectrode gaps in the range of 1.6-2.0 V. (Table 7.1), as a result of which their rapid heating occurred.

When the E1 source was switched on, the main discharge between the cathodes and anode was ignited and a significant increase in arc glow was observed.

The electric arc burned steadily in different modes. Arc burning was clearly observed not only from the holes on the cathode face, but also from the working surfaces of the cathodes. The maximum temperature of the cathode at the end reached 1600 K, and the anode 850 K.

Table 7.1 - Results of studies of cathode assembly with electric arc heating

№	t (sec.)	I∏K [A]		inagr [B].		1load [A]		U∏K [B]	
		nach.	con.	nach.	con.	nach.	con.	nach.	con.
Study 5.1. TCS=600K; $P_{ПРЕД}$=5'10^{-2} mmHg; 1=3мм; 1 =2мм$_{доп}$									
1	120	0	0	6,0	6,0	110	110	2,2	2,2
2	60	150	150	8,0	8,0	120	120	2,2	2,2
3	50	0	0	6,0	6,0	120	120	2,0	2,0
4	70	200	200	8,5	8,5	110	110	1,2	1,2
Study 5 2. TCS=600K; PPRE=5'10^{-2} mmHg; 1=3мм; ⅛∏=2мм									
1	60	0	0	5,8	5,8	125	125	3,1	3,1
2	19	60	60	7,9	7,9	125	125	2,0	2,0
3	30	140	140	7,9	7,9	125	125	1,5	1,8
4	10	100	100	6,8	6,8	125	125	1,8	1,7
5	14	80	80	6,5	6,5	125	125	1,8	1,9
6	55	200	200	7,9	7,9	150	150	1,5	2,0

As can be seen from the table, the voltage drop across the contact arc varied from 1.6V to 2.0V as the current flowing in the arc increased from 120A to 350A

The thermionic surface of one cathode was assumed to be equal to the sum of the cross-sectional areas of the holes, which was 0.03 cm^2 , and the total cathode assembly was 0.1 cm^2 . The cathode end area was 2.5 cm^2 , and the total cathode assembly was 7.5 cm^2 .

Moreover, the value of the total current in the electric arc between the cathodes and anode increased from 110 A to 260 - 300 A, and the voltage drop in it was 1.5 - 2.0 V. The discharge burned steadily for 3 minutes.

After 430 sec of the test, the arc began to burn erratically and the test was terminated. The study was terminated due to the possibility of anode melting. After investigation it turned out that the CHP parts were in good condition.

However, there were small meltings on the anode surface where electric arcs burned from the cathode holes.

In Studies 5.3- 5.4, a cathode assembly with three cathodes was installed in preparation for testing. The gap between the cathodes and the anode was 3 mm, and the gap between the heating cathodes (HC) and the main cathodes was 2 mm.

At first, the cathode heating circuit by contactors K_3 and K_2 was connected to the heating source E2 (Fig. 7.2) and when current flowed through the heating cathodes, they were heated up.

After closing the vacuum valve, cutting off the working volume of the product from the vacuum system, the ampoule with cesium was broken" and, cesium vapors entered the cavities of the cathodes and the area of the electric arc contact.

As in the previous study, after the cathodes were warmed up, contactor K1 was switched on and an electric arc was ignited between the cathodes and anode. During 100 - 200 s the arc burned unsteadily from different places of the cathodes. After K3 was switched off, the electric arc burned from all holes on the cathode end. There was a rapid heating of the working surfaces of the cathode. When the contactor K_1 was switched on, the main discharge was ignited and a significant increase of the arc glow was observed. The electric arc burned steadily in different modes. The arc burning from the holes on the cathode end was clearly observed. The maximum temperature of the cathode at the cathode end reached 1600 K, anode - 850 K.

After the eighteenth mode, when the test duration was 430sec, the arc began to burn unsteadily and the study was terminated.

After testing, all parts of the cathode and anode were in good condition with no signs of melting.

As can be seen from Fig. 7.4 and Fig. 7.5 the voltage drop in the contact arc varied from I,B to 4.8 V with increasing current flowing in the arc from 120 to 430A The calculated density in the contact arc varied in the range of 120 - 430 A/cm^2 .

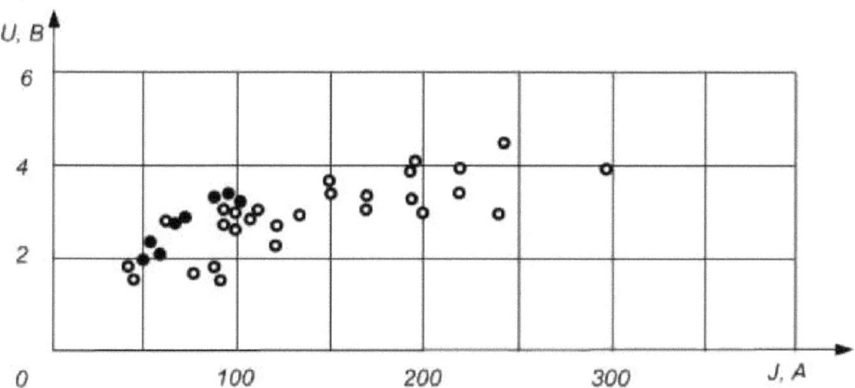

Figure7.4 - Voltampere characteristic of the electric arc of the plasma contact in studies 5.3.-5.8.

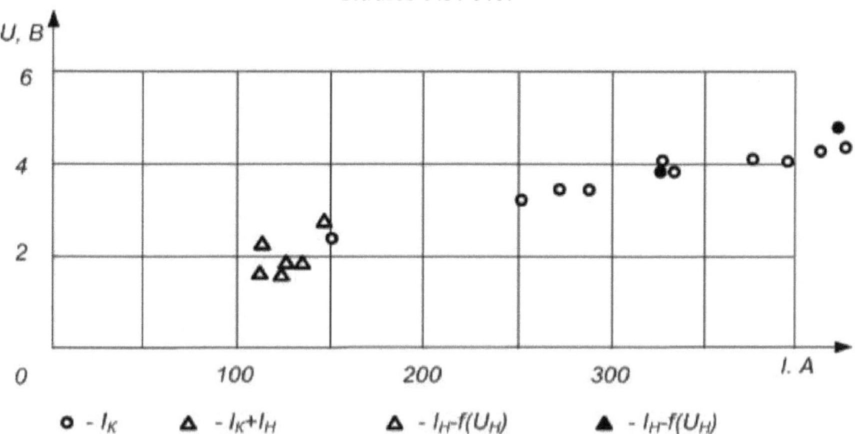

Figure 7.5 - Voltampere characteristic of the electric arc of the plasma contact in Study 5.3.

In this case, the thermoemissive surface of the cathode was taken equal to the sum of the areas of the calibrated cross-sections of the holes

In Studies 5.3 and 5.4, the electric arc burned stably only from the end holes both when contactor K2 and K1 were switched on.

In Study 5.5, the gap between the cathodes and anode was 3 mm, and the gap between the heating cathodes (HCs) and the main cathodes was 2 mm.

At first, studies were carried out when the contactor E1 was switched on, i.e. the characteristics of the main discharge were taken. However, the discharge pulsations, instability of its burning and high voltage drop in the contact arc were observed.

After switching on the electric arc heating, the discharge burning was stabilized, the value of the current at the switched on source E_2 , varied from 60 A to 200 A, and the voltage from 7.8 to 8 V.

The magnitude of the current when the source E_1 , was turned on, varied from 60 A to 200 A, and the voltage from 2.0 V to 5.0 V.

As can be seen from the WAC presented in Fig. 7.6 in this study, there is an insignificant dependence of the voltage drop in the contact arc on the magnitude of the electric current in the heating arc.

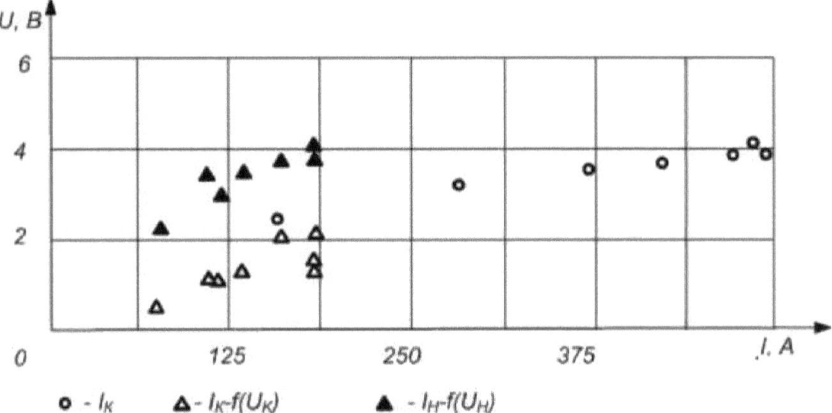

Figure 7.6 - Voltampere characteristic of the electric arc of the plasma contact in Study 5.4.

As can be seen from this characteristic at a current up to 200 A, when the current density in the discharge does not exceed 320 A/cm^2 , the voltage drop in the contact in the presence of an electric heating arc is less than with a heating arc and amounted to 1.6 + 2.4 V. At higher current values, the electric heating arc had no effect on the magnitude of the voltage drop in the contact.

The electric arc between the cathode and anode burned steadily and, as in the first test, the electric arc was observed from the cathode ends and from the holes on the cathode end.

The magnitude of the current when the source E_1 , was turned on, varied from 60 A to 200 A, and the voltage from 2.0 V to 5.0 V.

In studies 5.6 and 5.7, as in the previous studies, the gap between the cathodes and anode was 3mm, and the gap between the heating cathodes and the main cathodes was 2mm.

After applying voltage to the electrodes, the electric arc did not burn steadily, however, starting from the sixth seventh mode. i.e. after 230 sec burned steadily from the end surfaces and holes, both in the presence of electric arc heating and without it.

The test results are presented in the form of VAC in Fig. 7.7. As can be seen from the test results, the voltage drop in the contact arc was 1.8V to 3.5V when the current varied from 120A to 430A. In the presence of heating arc, the voltage drop in the contact arc was slightly less, indicating the positive effect of heating arc on the contact arc.

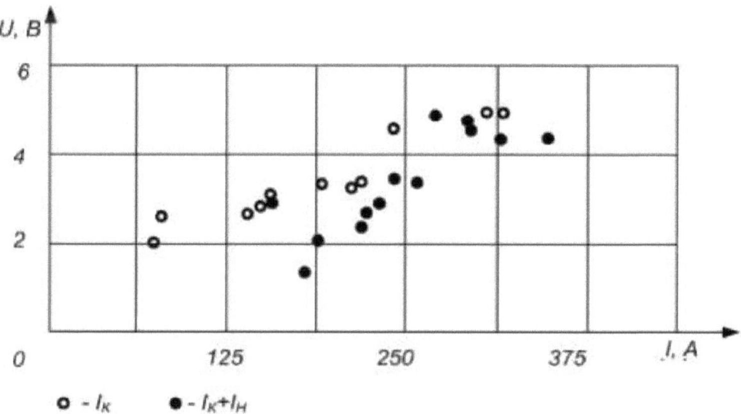

Figure 7.7 - Voltampere characteristic of the electric arc of the plasma contact in Study 5.5.

The nature of the flow of the electric arc combustion modes can also be seen from Figs. 7.8-7.9, where excerpts from oscillograms of studies 6.-6.7 are presented

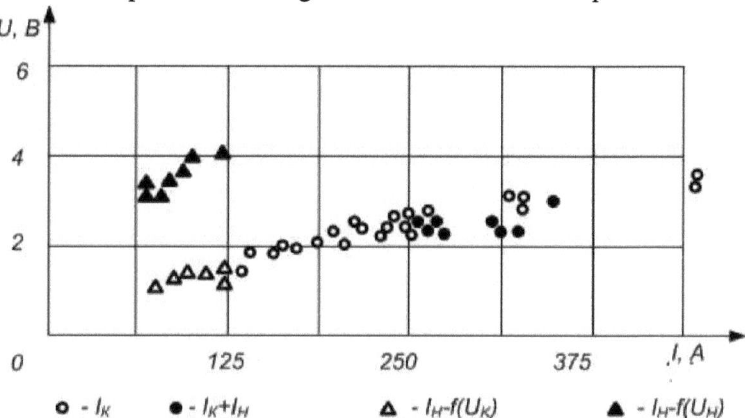

Figure 7.8 - Voltampere characteristic of electric arc of plasma contact in study 5.6

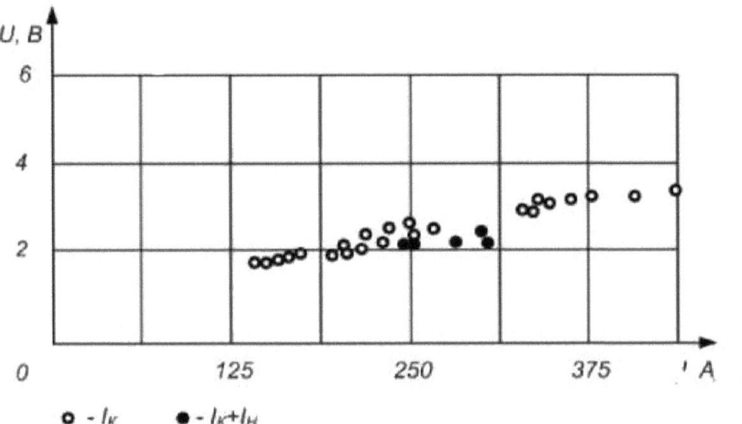

Figure 7.9 - Voltampere characteristic of electric arc of plasma contact in study 5.7

As can be seen from the results of studies 5.6 and 5.7. ,(Fig. 7.8 and Fig. 7.9), the voltage drop in the contact arc was 1.8V -3.5V when the current varied from 120A to 430A.

In the presence of a heating arc, the voltage drop in the contact arc was somewhat smaller, indicating a positive effect of the heating arc on the contact arc. The nature of the flow of the electric arc burning modes can be seen from Fig. 7.8 and Fig. 7.9, where excerpts from the oscillograms of the studies are presented.

In the presence of the heating arc, the voltage drop in the contact arc was slightly smaller, indicating a positive effect of the heating arc on the contact arc.

8. Investigations of electrical contact with a multi-cavity cathode assembly

Due to the satisfactory performance of the annular current collector investigated in the previous works, it was decided to start building and testing individual cathode assemblies.

The studies were carried out on the unipolar motor stand Fig. 8.1, which shows the moment of preparation of experimental studies.

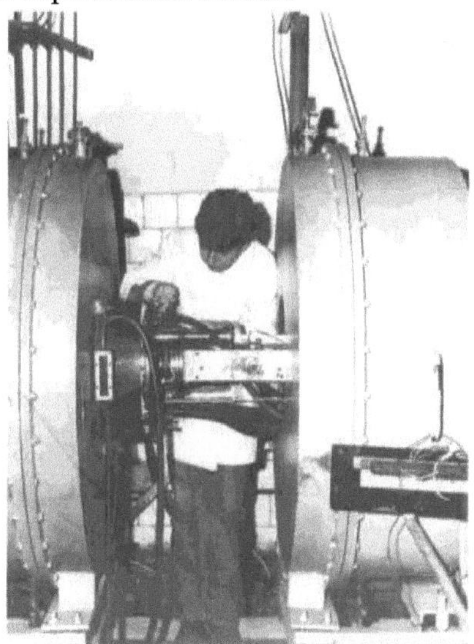

Figure 8.1 - Test bench for unipolar motor research

In the research, it was supposed to check the performance of the developed multi-cavity cathodes and the voltampere characteristics of the electric arc contact in different modes.

After that, it is obviously advisable to carry out studies of the most optimal designs of cathode assemblies, allowing to obtain the minimum voltage drop in the arc contact at a current density in the contact, equal to 100-200 A/cm^2 .

Studies 6.1-6.3 These studies utilized the cathode assembly of Fig. 8.2.

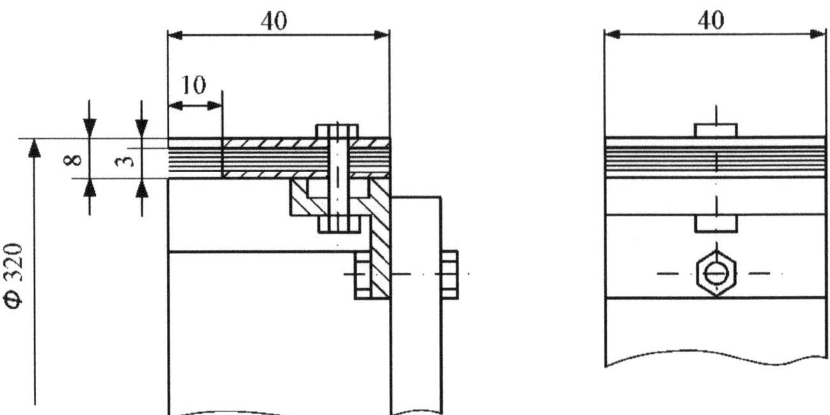

Figure 8.2 - Design of a multi-cavity cathode assembly.

The cathodes of this current-carrying device were made of 10 tungsten plates 0.3 mm thick, 40 mm in size, compressed into a package between two plates made of X18N10T steel, The thermoemissive surface of such a cathode was 1.2 cm^2 . The cathode assembly consisted of two sectors, on each of which three multi-cavity cathodes were fixed.

The thermal emissive surface area of the node was 7.2cm^2

After heating the unipolar motor body to a temperature of 570 K and the presence of cesium vapor in the working volume between the electrodes, after applying voltage to them, an electric arc of plasma contact appeared. In the initial period during (10-30s) in the observation window pulsations and breakdowns of the electric arc were observed. Further its burning was stable.

Moreover, when the electric current is reduced to 100-150 A, at current density equal to 15 -20A/cm^2 , the electric arc burned only from the end surfaces of the cathodes. With increasing electric current, the arc gradually spread to the side surfaces of the cathodes and at 500 - 600A it was observed both from the end and side surfaces. In addition, the presence of electric arc in the cavities of the cathodes was clearly seen.

With the increase of electric current up to 650A there was an increase of voltage in the electric arc of contact and its unstable burning The results of the test are given in the table and presented in Fig.-.

In Study 6.4, three multi-cavity cathodes were arranged on the current-supplying electrode.

After applying voltage to the main electrodes, the arc was ignited between them. In the beginning, the voltage drop in the contact arc reached 10-11V during 70sec. Starting from the time of 100 sec of the third mode, the burning of the arc stabilized. and the voltage in the arc was 4.5-60V at a current of 100-500A, In the beginning it was observed only from three cathodes, and then gradually spread to other cathodes. Moreover, as in previous tests, the change in the magnitude of the electric current led to a proportional change in the voltage in the arc, as well as in the size of the surface from which the arc burned.

Inclusion of the longitudinal magnetic field and its increase led to the fact that the side surfaces of cathodes, from which the electric arc burned, decreased and they shifted to the cathode ends. When the magnetic field strength reached the value of 1300 E at current density from the cathode surfaces 75 -100 A/cm^2 discharge burned only in the interelectrode gap between the cathodes and the disk.

The voltage drop in the arc with increasing magnetic field strength increased from 5.2V at H=400E, to 6V at H=1150-1300E. In addition, as visual observations showed, the longitudinal magnetic field stabilized the process of arc combustion.

After 220sec of arc burning, the voltage drop in the PC reached 11-12V at a current of less than 200A and therefore the test was terminated.

In Study 6.5, six multi-cavity cathodes attached to two sectors were used.

In this study, a probe made of tungsten wire with a diameter of 0.3 mm and placed in the hole of one of the cathodes so that it was at the level of the end of the cathode was used. The use of the probe made it possible to determine the voltage drop in the arc directly between the anode and the cathode and$_{зонд}$.

Starting from 80sec, the electric arc burned steadily from all cathodes. Moreover, with the increase of electric current up to 200-300 A at a density of 60-85 A/cm^2 arc was observed not only in the cavities of cathodes, but also from their side surfaces of cathodes.

The test results are presented in Table 8.1 and Fig. 8.3.

Table 8.1 - Excerpts from the oscillogram of study 6.5.

№	I1(A)	I2 (A)	U1(B)	U2 (V)	U1(3OHə)	Szcząd)	H(e)	T(sec)	L(MM)
Study 6. 5 P_{NACH} = 10^2 mmHg; P_{GOR} = 10^1 mmHg; S_K = 7 cm^2; T_{CS} = 600K; $T_{C.NACH}$ = 670K; $T_{D.NACH}$ = 700K									
1.	300	300	4,2	4,2	1,9	1,9		20	4
2.	200	200	3,5	3,5	1,6	1,6		20	4
3.	250	250	4,5	4,5	2,12	2,12		20	4
4.	450	450	8,0	8,0	4,72	4,72		20	4
5.	300	300	4,5	4,5	2,8	2,8		20	4
6.	400	400	4,5	4,5	2,04	2,04		20	4
7.	250	250	5,1	5,1	2,29	2,29		20	4
8.	350	350	4,5	4,5	2,04	2,04		20	4
9.	250	250	5,5	5,5	1,93	1,93	400	20	4
10.	250	250	5,5	5,5	1,50	1,50	800	20	4
11.	300	300	5,0	5,0	0,098	0,098	1150	20	4
12.	300	300	5,0	5,0	1,33	1,33	800	20	4
13.	350	350	4,5	4,5	1,78	1,78	400	20	4
14.	400	400	4,2	4,2	2,54	2,54		20	4
15.	650	650	6,0	6,0	4,3	4,3		20	4
16.	700	700	7,1	7,1	5,06	5,06		20	4
17.	750	750	8,0	8,0	6,0	6,0		20	4
18.	200	200	4,5	4,5	1,58	1,58		20	4
19.	800	800	7,5	7,5	6,0	6,0		20	4
20.	200	200	5,0	5,0	1,41	1,41		20	4
21.	150	150	4,5	4,5	1,33	1,33		20	4
22.	250	250	4,0	4,0	1,42	1,42		20	4
23.	350	350	4,4	4,4	1,73	1,73		20	4
24.	450	450	4,7	4,7	2,63	2,63		20	4
25.	550	550	5,5	5,5	3,53	3,53		20	4
26.	650	650	6,0	6,0	4,13	4,13		20	4
27.	700	700	7,2	7,2	4,7	4,7		20	4
28.	150	150	4,5	4,5	1,26	1,26		20	4
29.	200	200	3,6	3,6	1,24	1,24		20	4
30.	300	300	3,8	3,8				20	4
31.	400	400	4,0	4,0				20	4
32.	600	600	4,9	4,9				20	4

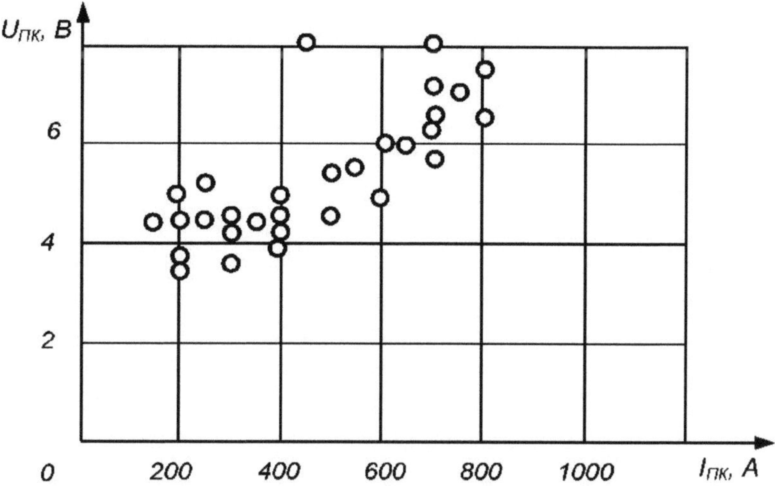

Figure 8.3 - Voltampere characteristic of the Study PC 6.5.

As can be seen from Fig. 8.3 and Fig. 8.4 with increasing electric current, the voltage drop in the PC arc increased, albeit slightly.

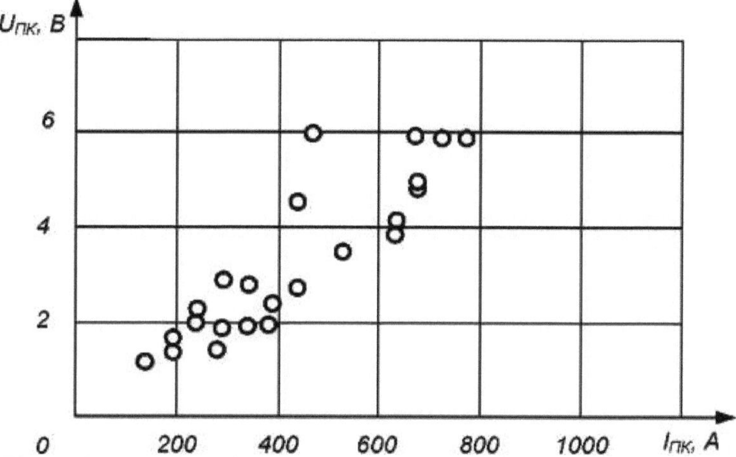

Figure 8.4 - Probe voltampere characteristic of the PC of study 6.5.

Thus, if at a current of 200 A the voltage drop in the contact arc was 4 V, then with increasing the current up to 800 A, i.e. at a density of 200A/cm^2 , the voltage in the contact arc increased to 6 V.

Incorporating a longitudinal magnetic field and varying its strength from 0 to 1150 E had very little effect on the arc voltage drop, although it did result in an increase in the contact arc voltage drop.

However, the same magnetic field resulted in a marked decrease in arc voltage at the same current values.

As in previous studies, the magnetic field significantly influenced the burning character of the electric arc.
As the magnetic field increased, the area of the side surfaces from which the discharge was observed decreased, and at H⁼ 1150E the electric arc burned only in the cathode cavities. Moreover, the burning of the electric arc became more stable and steady.
In Study 6.6, six multi-cavity cathodes attached to two sectors were used, the interelectrode gap between the anode and cathodes was 2.5mm
To measure the voltage drop directly in the contact arc, a probe inserted into the opening of the cathode was used. The voltage in the contact arc was measured using two conductors inserted into the working volume and attached to one of the sectors of the cathode and to the disk on the opposite side of the cathode in the arc burning zone. After applying voltage to the electrodes, an arc appeared between the cathodes and anode. But during 2 -2.5 min, it burned unsteadily, moving from one cathode to another.
Starting from the third mode from the sixth mode, after 120 seconds the character of the arc burning became stable, and gradually it spread to all cathodes, capturing both end and side surfaces. With the change of electric current in proportion to its value the surface of the arc burning changed, at a current of 100-250 A it was observed only from the ends of the cathodes, filling at the same time their cavities.
The arc voltage of the electrical contact, $U_{зон}$ ⅛ varied mainly from 1.5V to 5.0V with increasing current in the range of 200- 540A
If we compare the dependence of $U_з$ on IпК in tests 6.5 and 6.6, we can see a rather good coincidence of the results of these tests, and therefore we can assume that this dependence in test 5.6 should slightly differ from the similar dependences in the previous test. Moreover, the increase in the strength of the longitudinal magnetic field also led to a slight decrease in $U_з$

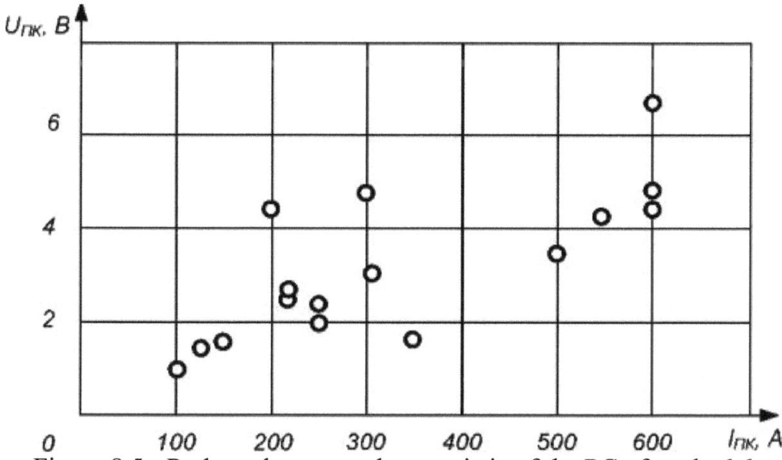

Figure 8.5 - Probe voltampere characteristic of the PC of study 6.6

In addition, the magnetic field promoted a more stable burning of the electric arc and reduced its burning surface.

After the test, the electrodes were in good condition.

Studies 6.7 and 6.8 were carried out using a cathode assembly having six multi-cavity cathodes fixed on two sectors, the thermoemissive surface of the cathode assembly was 3.5cm^2 , the interelectrode gap between the anode and cathodes was 2.5mm.

In Studies 6.7-6.8, a steady discharge from the cathodes was observed immediately after applying voltage to the main electrodes.

And at first the electric arc burned only from the two outermost cathodes located on the lower sector.

There was no discharge from the middle cathode. Later it gradually spread to other cathodes and burned from all cathodes visible in the observation window.

The voltage drop in the electric arc was 5-7 V at a current of 150300 A (Table. With increasing current more than 300 A Uпк reached 13 V, in connection with which the test was discontinued.

After disassembly of the experimental equipment and inspection of the cathode assembly and disk, brightly charged traces of arc burning from the two lower sectors and two lateral sectors were revealed on the disk working surface. The disk in the zone of arc burning from the two lower cathodes has insignificant melting.

Study 6.8 was carried out using a cathode assembly having six multi-cavity cathodes fixed on two sectors, the interelectrode gap between the anode and cathodes was 3.0mm.

During the study, at the initial moment of applying voltage to the electrodes, the discharge was running, not stable. However, starting from the mode 3-4,120sec, stable burning of electric arc was established. And it was the most intensive only from the lateral and upper sectors.

On repeated activations, the electric arc was ignited on the same cathodes.

When the magnetic field was switched on, the change of the arc burning surface proportional to the intensity H was clearly seen.

The voltage drop in the contact arc was 3.0-6.0 V with an electric current variation of 200-700A.

However, the magnetic field influenced the discharge characteristics leading to an increase in $U_{\pi\kappa}$ at the same current value.

In this study, the cathode temperature was recorded. The maximum cathode temperature of 1100-1200 K was set in the during 50-60sec and changed only by 100-150 K during the test.

Studies 6. 9 and 6.10 used a CG recruited from six sectors. The thermoemissive surface of the cathode assembly was12cm^2 , + The interelectrode gap between the anode and cathodes was 3.0mm.

During the initial period of the test, the discharge was unstable. Later, the electric arc burned steadily and at 350 A was visible on the entire side and end surfaces. With the increase of electric current more than 800 -1000 A, an increase in the thickness of the plasma layer on the side surfaces of the cathodes was observed. The value of the electric current in the ninth study varied from 200A to 1900A at a voltage drop in the contact arc of 6.0-10V, Table 8.2.

Inclusion of the magnetic field, as before, led to the pulling down of the discharge to the end surfaces of the cathode. However, if at a current of 180-200 A the electric arc burned only from the cathode face, then with increasing the electric current up to 1200-1400 A the arc burned in some part of the cathode side surfaces.

At the same time, both without and with the magnetic field, the intensity of discharge burning along the ring was somewhat different.

Increasing the magnetic field from 0 to 1300 E had an insignificant effect on UBK, decreasing its magnitude. Moreover, as the current in the plasma contact decreases, the influence of the magnetic field on the voltage drop in the arc decreases. After the tests, the cathode assembly and anode were in satisfactory condition.

Table 8.2 - Excerpts from the oscillogram of study 6.9

№	I .nach (A)	I.KOH. (A)	U.ui···ι^ι ι (B)	i.con (B).	6	7	H (э)	t (sec)	l (mm)
PNAc = 10^{-2} mmHg; PГОР = 10^{-1} mmHg; SK = 22 cm^2; TCS = 550K; TC.NAc = 650K; TDNACH = 670K									
1.	600	400	6,0	7,0				37	3
2.	350	180	7,5	9,5				50	3
3.	200	180	9,0	8,5			0-1300	15	3
4.	200	180	8,5	8,7			0-1300	37	3
5.	180	200	8,7	8,5			1300-0	5	3
6.	1550	1600	10,0	10,0			400	12	3
7.	1300	1350	9,0	8,0				30	3
8.	1400	1450	8,5	7,5			0-1300	5	3
9.	1450	1400	7,5	7,5			1300	10	3
10.	1900	1900	9,0	9,8			1300	13	3
11.	800	950	7,0	6,0			1300	15	3

In the tenth study, the electric current varied from 550A to 1450A at voltages of 6.5V to 7.0V. Inclusion of the magnetic field, as before, led to the shrinking of the discharge to the end surfaces of the cathode. However, if at a current of 180-200 A the electric arc burned only from the cathode face, then with increasing the electric current up to 1200-1400 A the arc burned in some part of the cathode side surfaces.

At the same time, both without and with the magnetic field, the intensity of discharge burning along the ring was somewhat different.

Increasing the magnetic field from 0 to 1300 E had an insignificant effect on UBK. reducing its value. Moreover, the influence of the magnetic field on the voltage drop in the arc decreases with decreasing t oka in the plasma contact.

Conclusions.

A cumulative analysis of the results of all the tests performed leads to the following conclusions:

1. the use of multi-cavity cathodes under appropriate conditions (Рцз = 5-6 tor; U PC= 1,5-2 mm; Pprev. $=10^{-2}$ tor. allows to create a cathode unit, providing an electric current equal to several thousand amperes at a voltage drop in the contact arc of 5-6 V,

1 .Significant variation of plasma contact arc parameters, unstable character of arc combustion and cathode unit steady-state operation is caused by non-uniform density of cesium vapor in the interelectrode gap of the current collector.

3 .Increasing the current density in the arc more than 90-100 A/cm^2 was limited both by the underestimated and non-uniform cesium concentration in the PC zone and by the inefficiency at the investigated modes of heating the working surface of the cathode by the current-carrying arc.

4 .Changing the interelectrode spacing within 5-8 mm at the induced cesium feeding scheme and magnetic field strength up to 1300 ersted had no significant effect on the PC arc parameters.

5 .Since the rotation speed of the disk has no effect on the characteristics of the contact arc, the research should be carried out with fixed electrodes (anode, cathode,).

6 .The thermoemissive surface of the cathode should be made of porous tungsten for the passage of cesium vapor into the contact arc zone.

Literature.

When writing the book, the STOs published in the Kremenchug branch of the KHPI in 1982-1988 were used.

Printed by Books on Demand GmbH, Norderstedt / Germany